I Can't Do This By My Cell-f
A Journey Through MITOSIS

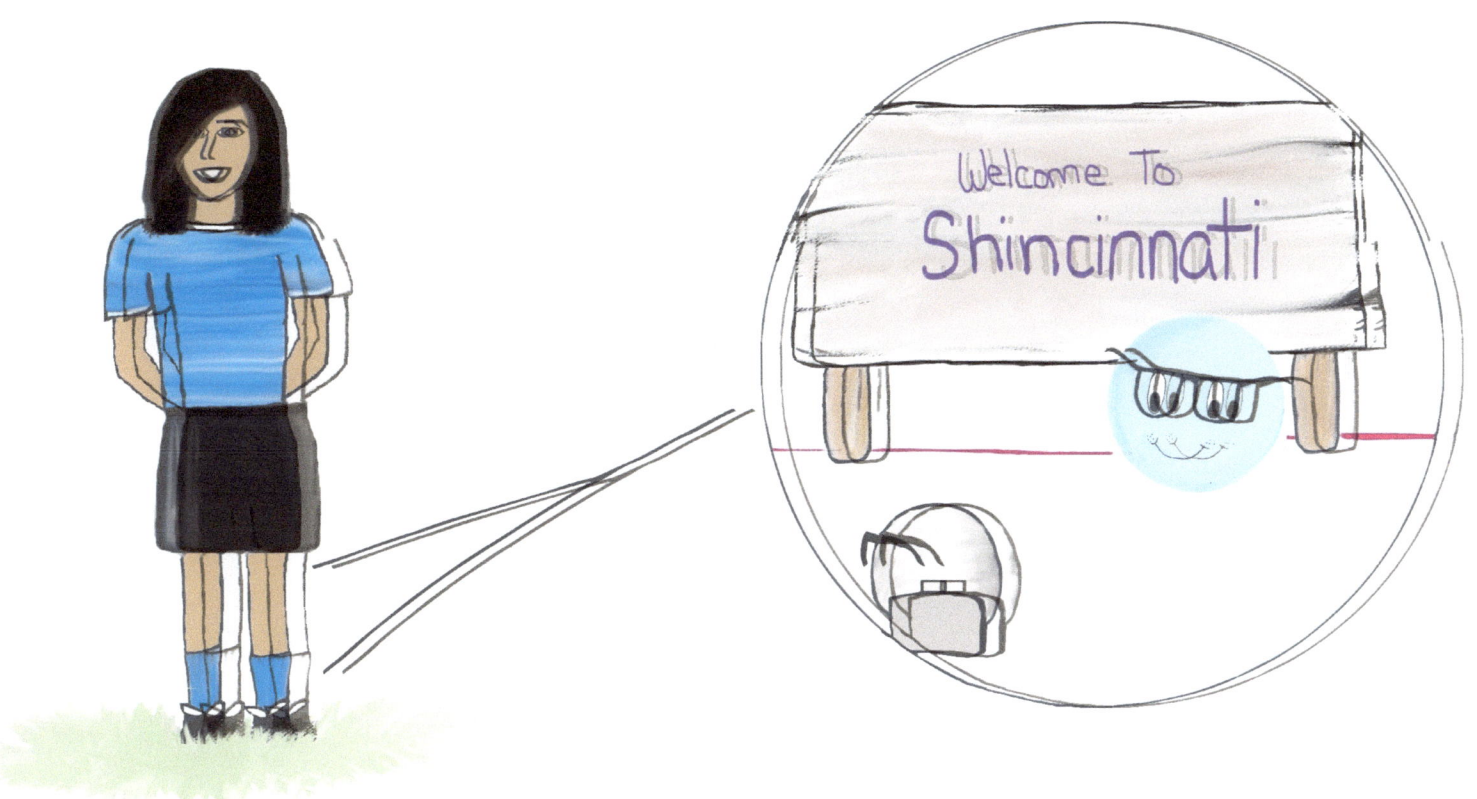

It was just another night for Celline who was out and about, making sure all was well in her town. You see, Celline lived in Shincinnati, one of the biggest cities in Hannah. Hannah was the adventurous, clumsy, funny and sweet host where many cells and organisms lived to help protect her against villainous viruses or catastrophic cuts.

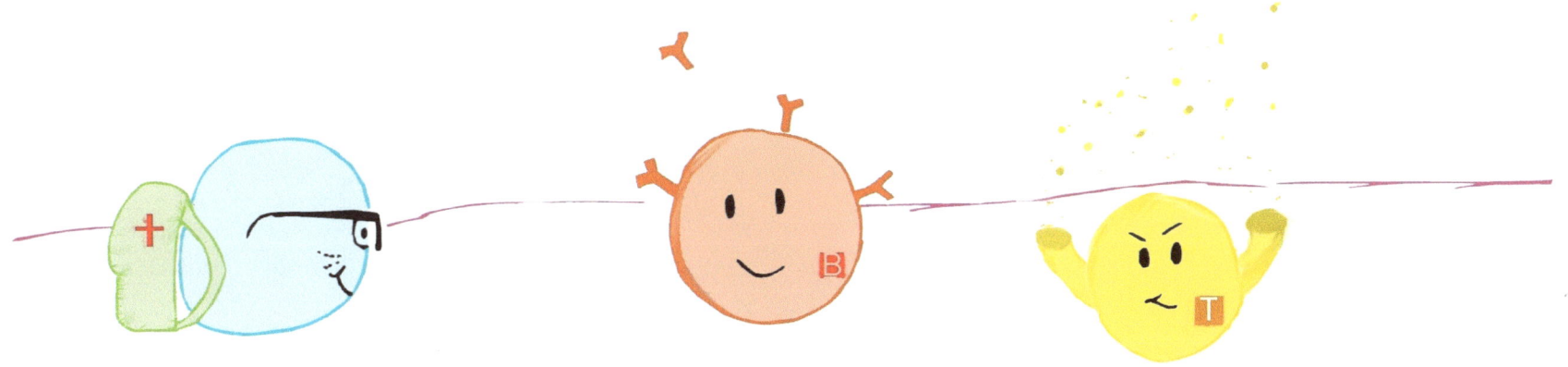

Everyone who lived in Shincinnati City was very busy. Celline had been living in Shincinnati for quite some time now. She was now the head of the emergency unit, and she loved her job! She was able to travel all around the body to talk to several cells, such as the B-Cells, who release antibodies, and T-cells, who destroy any unknown or dangerous cells. She thought they were very cool!

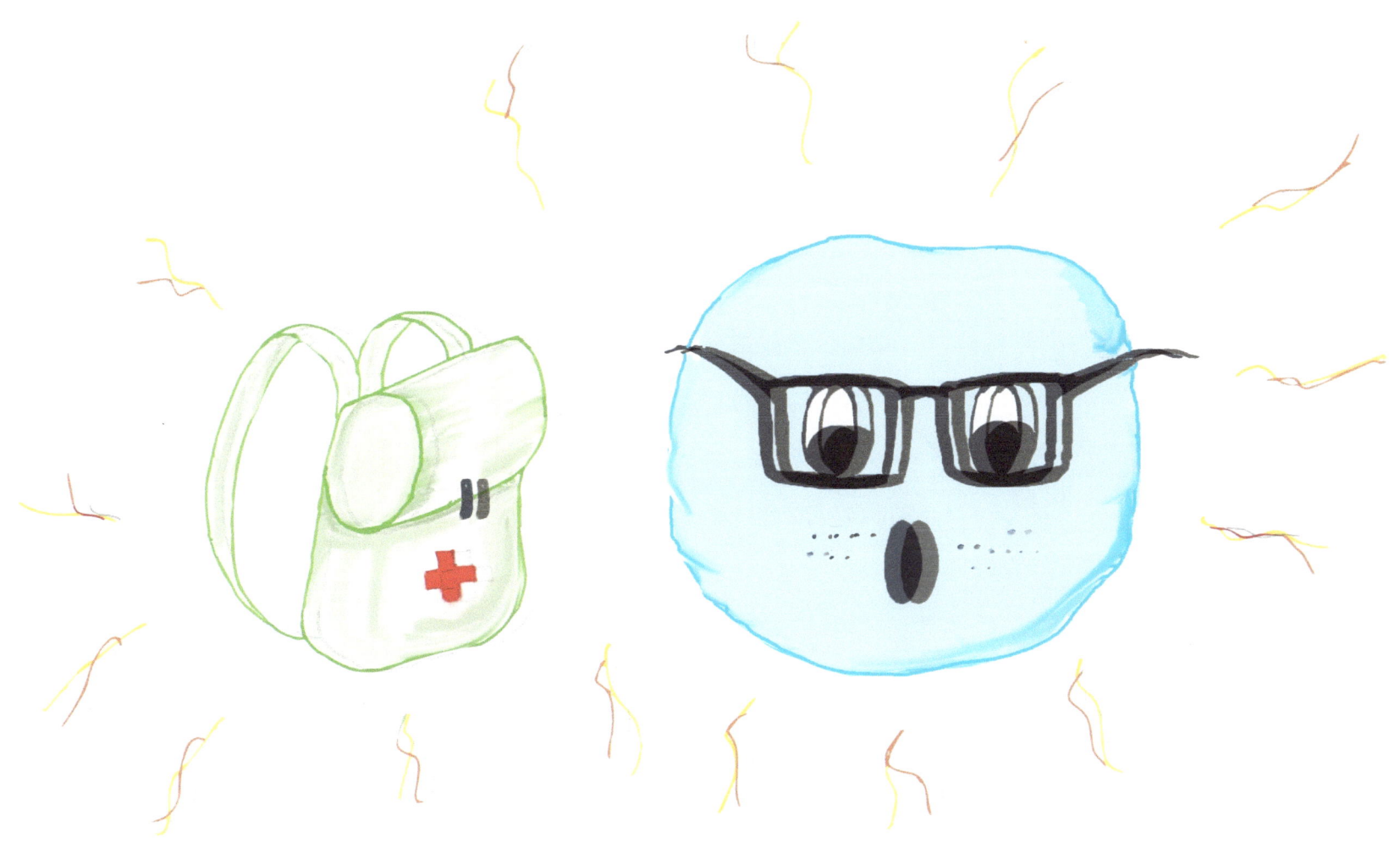

Celline was equipped for any kind of emergency... Or was she?
In the middle of one of her daily runs through the bloodstream, Celline felt a sudden jolt. All of a sudden, Hannah was beginning to grow! Celline was not sure what to do. Was she supposed to run to the city hall? Should she hide until it was over? This was the dreaded growth spurt. Only Celline had known it would come someday, but it still caught her off guard.

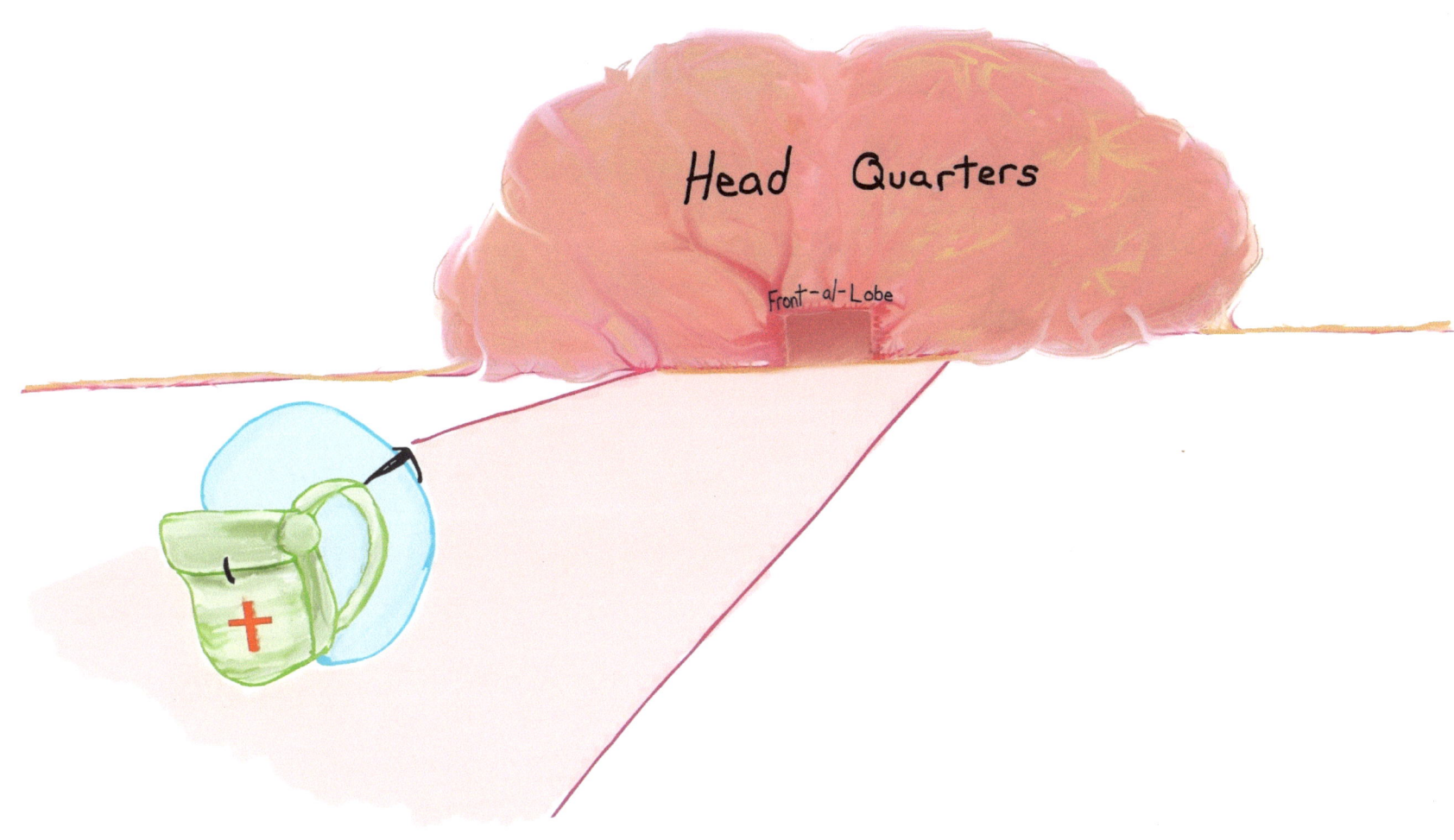

As the head of the emergency unit she knew she could not patrol everyone by herself, so Celline ran to Head-Quarters and asked the lobe-iest to call for more help.

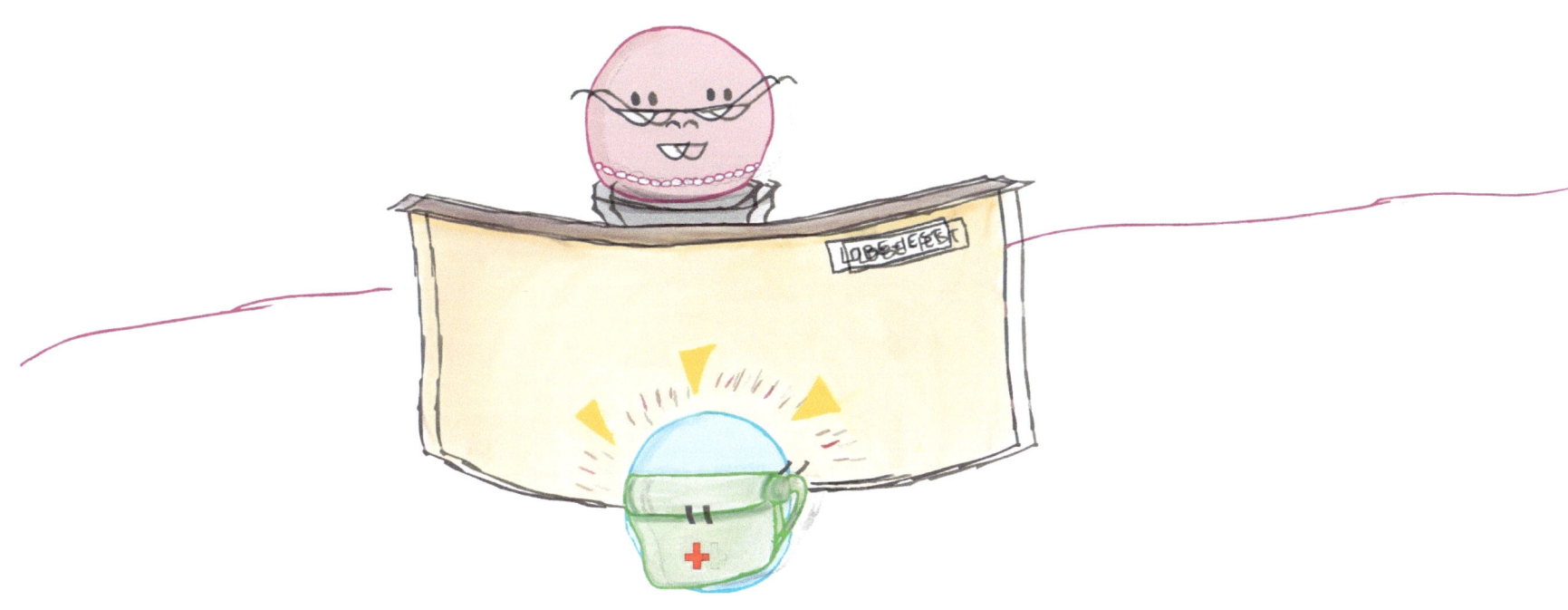

"All the help is busy, this has grown into a larger issue than we first anticipated," said the lobeiest.

Celine was in a panic, "Then how will I get anyone to help me! I can't just recruit anyone! We've never dealt with this kind of situation before!"

The lobeiest sat thinking for a little while.

Finally she said, "Well, how about cloning yourself?"

"Clone?" Celine replied. "Yes! I could use some help from someone just like me! Tell me how to do it!"

"Well, before we begin, I'll just say this. It will take a lot out of you," the lobeiest said.

The lobe-iest then led Celline to the top chamber and handed Celline a book. It was called *Mitosis: i.P.M.A.T.C.* This was exactly the help that she needed.
Celline rushed back to Shincinnati and got right to work so she could patrol her city. She opened the book and began to read:

Step 1: Interphase

This is the MOST important part! It will take the longest amount of time. Check to make sure you are ready for such a big change. If you are not ready STOP READING.

If you have determined that you are ready you may begin replication. Replicate your interior organelles, but hold back on the DNA; keep them scrambled for now. After replication, you can begin the process of MITOSIS, and you will be on your way to cloning yourself. Good Luck!

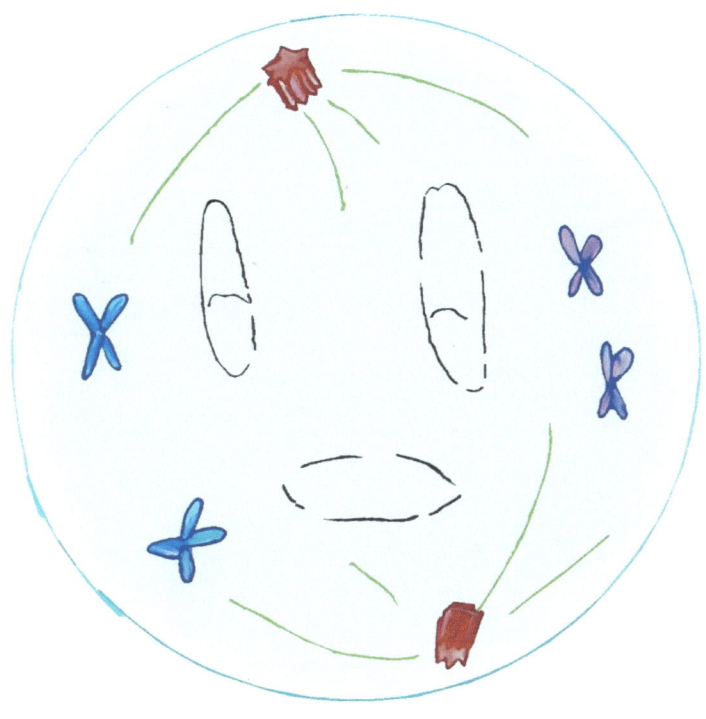

Step 2: Prophase

Now take your DNA and condense it into chromosomes. This will make your DNA visible and easier to handle. As you do this, your nucleus will disappear and your centrioles will drift to opposite ends. Don't panic! This is normal.

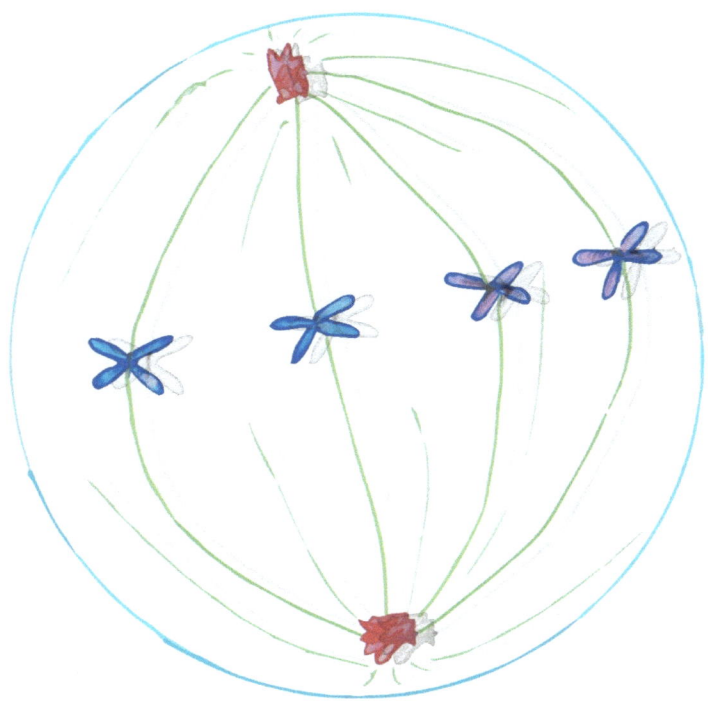

Step 3: Metaphase

Now that your DNA is visible, microtubules will buckle onto the kinetochores in order to move and line the chromosomes up in the middle of the metaphase plate. Keep them organized!!
You don't want to lose any!!

Step 4: Anaphase

CAUTION!
Be careful to do this step right!
If you want your clone to be identical to you, take heed and go slowly!
Take your chromosomes from the metaphase plate and begin to pull them apart to the opposite poles.

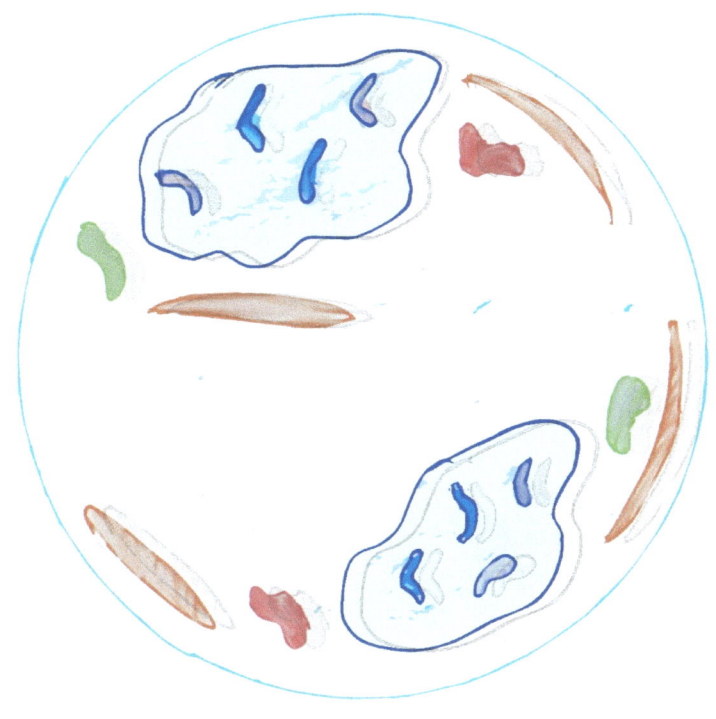

Step 5: Telophase

Well done. You have successfully duplicated within yourself! Now, it's time to truly become separated into identical cells. Begin to form a membrane around the two sets of chromosomes to keep them in one place on either side of the cell.

Step 6: Cytokinesis

Finally, you will begin to form a cleavage furrow -- that is, forming a cell membrane to distinguish the cells. This will result in two separate daughter cells!

Celine followed each step precisely. She did it! Now she had another cell just like her to help her out. However, just one was not enough. Celine cloned herself multiple times, following the same steps again and again. Finally she had enough helpers to help her with Hannah's growth spurt. Celine sent out her clones to handle the growth spurt and whenever they needed extra help, they would clone themselves and get the job done. Now Hannah's Emergency Unit was ready for anything that came their way!

www.ingramcontent.com/pod-product-compliance
Lightning Source LLC
Chambersburg PA
CBHW051942210526
45473CB00006B/2354